ELEMENTS OF THE
THEORY OF RESONANCE

ILLUSTRATED BY THE MOTION OF
A PENDULUM

By

E. W. BROWN, Sc.D., F.R.S.

Josiah W. Gibbs Professor of Mathematics
Yale University

CAMBRIDGE
AT THE UNIVERSITY PRESS
1932

CAMBRIDGE UNIVERSITY PRESS
Cambridge, New York, Melbourne, Madrid, Cape Town,
Singapore, São Paulo, Delhi, Tokyo, Mexico City

Cambridge University Press
The Edinburgh Building, Cambridge CB2 8RU, UK

Published in the United States of America by Cambridge University Press, New York

www.cambridge.org
Information on this title: www.cambridge.org/9781107659759

First published 1932
Re-issued 2011

A catalogue record for this publication is available from the British Library

ISBN 978-1-107-65975-9 Paperback

CONTENTS

PREFACE

This pamphlet contains an attempt to describe and analyse the phenomena which are peculiar to resonance in an elementary manner. Although the subject is of fundamental importance in many mechanical problems it has received but little attention in the text books. The treatment, which consists of an investigation of the changes which take place in the amplitude and phase of a vibrating element under certain types of forces, appears to be convenient both for the mathematical development and for the interpretation of the results.

The issue of The Rice Institute Pamphlet, Vol. xix, No. 1, in the present form, is made with the permission of the Trustees of the Institute and the co-operation of the Cambridge University Press, and is due to the hope that it may prove to be useful to students and others who may not have easy access to the publications of the Institute. My thanks are due to Professor C. G. Darwin for his helpful criticisms, especially in the mode of presentation of a subject in which the method of approach and certain of the results appear to have some degree of novelty.

ERNEST W. BROWN

April 1932

ELEMENTS OF THE THEORY OF RESONANCE ILLUSTRATED BY THE MOTION OF A PENDULUM[1]

INTRODUCTION

1. The original observations which gave rise to the word "resonance" were audible sounds which are familiar to most of us. A note is struck on a piano or other musical instrument and some body—a wall or another musical instrument —will take up the same note and will continue to sound it even after the original source is stopped. The megaphone takes up the vibrations of the air produced by the voice and gives them out again, concentrated in a particular direction. A vibrating body, for example, the cylinders in the engine of an automobile, will set up vibrations in other parts of the car, certain vibrations being noticeable at certain speeds and others at other speeds. To all these phenomena the term resonance is applied.

But the actual nature of the phenomena is not always the same. The megaphone and loud speaker are designed to take up and emit any vibration within a certain range of frequency. On the other hand, the audible vibrations of a stretched wire are confined to a limited number of sounds as long as we keep the tension unaltered, and these bear definite relations to one another; to produce resonance in such a wire, it is necessary to sound a note with a frequency

[1]A course of lectures, given at the Rice Institute, April 22, 23, and 24, 1931, by Ernest William Brown, D.Sc., F.R.S., Josiah W. Gibbs Professor of Mathematics in Yale University

1

very near that of one of these modes of vibration. These illustrations bring up the necessity for a definition of what are usually called the *natural* frequencies or periods of vibration of any system.

2. A stretched wire when made to vibrate appears to give out the same note, in general, however it may be struck or when the bow of the violin is drawn across it. An ordinary drinking glass exhibits the same phenomenon: it gives out the same note when struck as when the finger is wetted and run around the top edge of the glass. The periods of these sounds are called "natural periods" of vibration: they depend only on the construction of the mechanism and not on the manner in which its vibrations are started.

But this last statement is not exact. Actually the note given by a vibrating wire is not quite the same when the sound is loud as when it is soft, although the difference is not easily detected even by a sensitive ear. In the former the excursions of the wire—usually called the amplitude of vibration—are greater than in the latter, and the period of vibration which gives the pitch of the note depends to some extent on this amplitude. The "natural" period is, in fact, a mathematical fiction since it is only present when the vibrations have infinitely small amplitudes, which amounts to saying that the wire is not vibrating at all. More properly a natural frequency should be defined as the lower limit of the frequency of that particular mode of vibration. It is necessary to insist on this change of frequency with change of amplitude because the existence of the *phenomena* of resonance depends on the existence of this change.

3. A mechanical system which is free to move at all can, however, usually be made to vibrate with any frequency whatever. When any such vibration is impressed on the system, it is called a forced vibration. It may be present

at the same time as the "natural" vibration so that the actual motion is compounded of the two types. It is usual to treat the frequency of the forced vibration as unchanging and in many problems this procedure can be justified, but this is not always the case. If two wires be tuned so as to give out notes with nearly the same pitch, each will force its frequency on the other and there will be a reaction of this effect which must sometimes be taken into account. The procedure fails altogether when the pitches of the two notes are sufficiently nearly the same.

4. The usual definition of resonance is the state of motion which is present when the natural and forced frequencies are the same, but it is evident from the remarks made above concerning the change of frequency with change of amplitude, that the definition lacks precision. It will be shown that the state of motion in which the frequencies are the same is fundamentally different from that in which they are not the same. Not only is it impossible to represent the two states by the same mathematical formula, but there is a fundamental physical discontinuity separating them. It is the exploitation of this discontinuity which is one of my chief objects in these lectures, partly because it is a feature of many mechanical problems in which resonance takes place and partly because I believe that it has far-reaching effects in gravitational problems and particularly in the past and future history of the solar system.

5. To illustrate the phenomena, consider two piano wires, one of which is being "tuned" so as to give the same note as the other. As the pitch of the second note approaches that of the first with the change in tension, the phenomenon known as "beats" occurs. Apparently, the two notes have the same pitch but the volume of sound rises and falls at regular intervals; this interval is called the "beat" frequency.

Actually the periods are not the same, but the ear is unable to distinguish the difference because its function is mainly to integrate the impulses which it receives. When the two vibrations have the same phases, the two sets of impulses are added together and increase the volume of sound; when the phases are opposite, the impulses nearly cancel one another and the sound is much diminished. The frequency of the "beat" is nothing else than the difference of the frequencies of the two notes.

As the tuning proceeds, the frequency of the beat becomes smaller, that is, the interval between the maxima of the sound becomes longer, until a stage is reached in which it can sometimes be heard and sometimes not. According to the theory, as developed in a later section, the audibility of the beat will depend on the magnitudes of the impulses given to the notes; the beat may be heard when the notes are struck gently but not when they are struck hard. With very slight further tuning, the beats disappear altogether, and we have the phenomenon known as resonance in which the two periods are exactly the same.

6. The last statement may give rise to the impression that to produce resonance, it is necessary to adjust the tuning with a degree of exactness which anyone familiar with experimental work would say cannot be achieved. It is not so. Resonance takes place when the two frequencies differ by a small but measurable amount. In fact, a certain degree of "tuning" can be made without destroying the resonance. What has actually happened is that we have produced a common average period of oscillation. About this common period, however, a new oscillation has been set up, and when we "tune" slightly, the alteration which takes place is one not in the common period but in the amplitude of this new oscillation. In astronomy, it is known as a physical "libra-

tion." The frequency of this oscillation is quite unconnected with the beat frequency; it depends mainly on the construction of and tension of the wires and on the magnitude of the blows which are given to set them vibrating. Sometimes it has a minimum period which is approached as the amplitude dies down. It is, however, doubtful whether it could be detected without the use of elaborate apparatus designed for the measurement of very small differences in the intensities of sounds.

7. Vibrating wires will not be used further for the illustration of the phenomena of resonance, partly because the oscillations are so rapid that observation is difficult, but mainly because of mathematical and physical difficulties which make the analysis very complicated.[1] We have, however, in the simple pendulum a vibrating system in which experiment can approach a simple mathematical theory very closely. The pendulum will be used in two ways. In later examples, it is an oscillating system on which various types of external forces can act. But its more important function here is due to the fact that the equation which gives the motion of the pendulum will be shown to be the same as the fundamental equation which we reach when considering resonance in a wide variety of types of mechanical systems. If then we have a complete solution of the motion of a pendulum, it is necessary only to reinterpret it for any case in which the same differential equation arises.

8. The pendulum in this latter case is, however, one which can perform complete revolutions as well as oscillate about a horizontal axis. A more practical form of illustration is that of a bicycle wheel loaded at one point of its rim by a weight clamped on; it is mounted on a fixed axis, so that it can turn freely in its bearings. Such a wheel can be made

[1] See section 44.

to oscillate about its position of stable equilibrium or can make complete revolutions in either sense according to the manner in which it is started, and the differential equation reduces at once to that of the pendulum, in the absence of frictional forces. It will be shown that the two types of motion—complete revolution or oscillation—correspond respectively to non-resonance and resonance conditions, and that the fundamental phenomena of the latter can be exhibited without difficulty. With the results for the simple pendulum in mind, the equations can be generalized and an approach made to more complicated problems such as those which are presented by the "problem of three bodies."

9. Certain mathematical features in the investigation of resonance should be mentioned at the outset. The most fundamental of these is due to the fact that *the phenomena cannot be investigated by linear differential equations alone.* In the ordinary theory of small oscillations, the periods are considered as independent of the amplitudes. This is an approximation which is usually sufficient as long as the amplitudes remain small. But the approximation is no longer valid when the amplitudes increase, and this usually happens when two of the frequencies in the system are nearly the same.

Take, for example, the equation

$$(9.1) \qquad \frac{d^2x}{dt^2} + n^2x = m \sin n't.$$

When $m = 0$, this gives a harmonic oscillation with frequency n. When m is not 0, there is, in addition, a similar oscillation with frequency n'. As long as n, n' are unequal we have the solution

$$(9.2) \qquad x = \lambda \sin(nt + \lambda_0) + \frac{m}{n^2 - n'^2} \sin n't,$$

where λ, λ_0 are arbitrary constants.

When, however, $n = n'$, the solution is

$$(9.3) \qquad x = \lambda \sin(n't + \lambda_0) - \tfrac{1}{2} \frac{mt}{n'} \cos n't.$$

This contains an oscillation with a coefficient which increases continually with t.

A common practice is the insertion of a "damping" factor $\mu dx/dt$, on the ground that all mechanical systems are subject to friction. It is easily seen that this prevents the coefficient from becoming infinitely great with t, but it will be shown below that it fails to give even an approximation to the motion of certain mechanical systems under resonance or near-resonance conditions. The frictional force acting on the pendulums used for geodetic surveys, for example, is so small that it may be neglected in comparison with certain types of disturbance which produce resonance, and the motion in the latter case is entirely different from that which would be produced by a frictional force. In the problem of three bodies, this device cannot be used; in most of the applications to the solar system, there is no evidence of any frictional effect.

10. The real defect lies in the assumption that the forces in a vibrating system are proportional to the displacements, so that the equations of motion are linear. This is not true in any mechanical system we know, although many systems approach it very closely. As long as no two of the frequencies in the system are very nearly the same, the assumption gives a good approximation to the motion, mainly because the amplitudes of the oscillations remain small. As soon, however, as the amplitude begins to become large, as it does when two of the frequencies are nearly the same, the approximation fails. It is necessary to take into account the effect of finite amplitude and this demands the use of non-linear equations.

Other cases common to many gravitational problems, are those in which the disturbing forces depend partly on the displacements of the system. Some examples of this will be given in which it will be shown not only that non-linear equations must be used, but that the methods of approximation usual in such cases fail. The nature of this failure will be easily seen by the mathematician when it is caused by the attempt to expand a function in positive integral powers of a certain constant instead of expansion in powers of the square root of this constant.

11. A feature of motion under resonance conditions already hinted at in section 6 will be brought out in detail by the examples given below. This feature, which cannot be too strongly emphasized, is that resonance is not a *single* special case of motion but is a *group of cases extending over a finite range of values of the constants*. It is true that one of the constants, previously arbitrary, is given a particular value; this constant, the frequency, becomes the same as that of the disturbing force (we shall show that another constant, the phase, must also have a special value).

Suppose, however, we regard this case as a particular solution of the equations of motion and then proceed to find the possible deviations from it. If the resonance motion is stable, we find that small oscillations about the resonance configuration are possible. In other words, we find that on the average the resonance relation is maintained, but that there are periodic deviations from it, and that the period of these deviations has no direct relation to the resonance period.

We find also that the amplitude and phase of these deviations are arbitrary constants of the solution. Thus though two arbitrary constants have been given special values when we set up the resonance relation, two new arbitraries appear in the deviations from this resonance relation. The chief period of oscillation is "locked" to that of the disturbing force, but the locking does not prevent deviations to and fro if the system be slightly disturbed.

The mathematical and physical characters of the motion are quite different when there is resonance, from those when resonance does not exist. Neither can be described in terms of the other.

12. These differences between the two groups of cases suggest that there must be at least a mathematical discontinuity separating them. This will be shown to be the actual fact. The differential equations for each group are the same, but the solutions of these equations are not continuous functions of the arbitrary constants. The formulae which give the motion have to be completely changed. At one place there will be found a solution which belongs neither to the one case nor to the other. The place corresponds to a discontinuity of an analytic function.

In a certain sense, this discontinuity is physical as well as mathematical. Suppose that we have been able to deduce the constants in a certain type of motion from observation. Usually small changes in the constants within the errors of observation produce only small changes in the subsequent calculated motion. At the point of discontinuity, however, such small changes in the observed values are found to change fundamentally the character of the subsequent motion. There is a certain degree of analogy between this case and that of equilibrium at an unstable position where the subsequent motion often depends on the nature of a minute and perhaps not measurable disturbance.

The quantities which we ordinarily associate with motion, namely, position, velocity, acceleration are continuous in the sense that there are no sudden changes in their values with a change in the time. The discontinuity is that of indeterminateness. Not only is there only a small change in them with a small change in the time, but the change may be still small with a finite change in the time, while the particle is passing through this particular stage. Later it may move in a manner which is easily calculable. Thus we might be unable to distinguish between the position at a given future time and that at a time later by a finite interval. In other words, we are unable to predict the position after a given interval of time.

13. The existence of this discontinuity in most cases. prevents calculation of the motion of a system if it passes from non-resonance to resonance conditions. Calculations are made of the motions as the system approaches resonance and also of the motion when the resonance is fully established. For the stages in between appeal is made to some simple physical case like that of the pendulum where experiment can give a qualitative description of the probable changes which take place. In the simple examples worked out below there will not be much doubt that this method of procedure gives the principal features of the motion.

For logical completeness, proofs of the existence and convergence of the series used should be given. Owing partly to limitations of space and partly to avoid the introduction of developments which are of more interest to the mathematician than to the physicist, these have been omitted. It is necessary to know the forms which the series must take for purposes of calculation; that these forms will give approximate numerical results is assumed on the basis of past experience, or is capable of proof by known methods which are not developed here.

I

THE SOLUTIONS OF THE EQUATION $\dfrac{d^2x}{dt^2}+\kappa^2\sin x=0$.

14. It has been mentioned above that the equation which in many cases gives the resonance phenomena is the same as that of the motion of a simple pendulum. Since this motion is easily visualized, it will assist in the comprehension of the phenomena if the solutions and their physical meanings are carried out in detail. Certain features of the motion of a pendulum, usually left aside, will be emphasized here on account of their importance in the later applications.

Let us suppose that a particle be attached to one end of a light (weightless) rod of length l, the other end being attached to a horizontal axis about which the rod can turn without friction in a vertical plane. If g be the acceleration produced by gravity and x the angle which the rod makes with a downward drawn vertical line at time t, it is well known that the motion is given by the differential equation

$$(14.1) \qquad \frac{d^2x}{dt^2}+\kappa^2\sin x=0,$$

where $\kappa^2=g/l$. According to the manner in which it is started, the rod may make complete revolutions or it may oscillate to and fro on each side of the vertical through any angle up to 180°.

The immediate object in view is the discovery of the different types of solution of (14.1), considered merely as a differential equation. The physical illustration aids in giving a concrete idea as to the nature of these solutions.

Simple Pendulum

The equation

$$\frac{d^2x}{dt^2} - \kappa^2 \sin x = 0$$

can be reduced to (14.1) by the substitution $x+\pi$ for x. A more general type,

$$\frac{d^2x}{dt^2} + f'(x) = 0,$$

will be discussed later.

15. On multiplication of (14.1) by $2dx/dt$, we can integrate and obtain

(15.1) $$\left(\frac{dx}{dt}\right)^2 = C + 2\kappa^2 \cos x,$$

where C is an arbitrary constant to be determined from the initial value of x and of the angular velocity dx/dt. The different types of solution depend on the value of C. Since $\cos x \geqq -1$, we must have $C \geqq -2\kappa^2$, in order that the velocity may be real.

There are three cases.

(i) $C > 2\kappa^2$. The velocity never vanishes and as it is always finite, it must be always positive or always negative. The rod is making complete revolutions clockwise or counterclockwise.

The integral of (15.1) is

(15.2) $$t + \text{const.} = \int \frac{dx}{(C + 2\kappa^2 \cos x)^{\frac{1}{2}}},$$

where the square root may have either sign. Since $|2\kappa^2/C| < 1$, the integrand may be expanded into a series of the form

(15.3) $$\frac{1}{n} + \sum_{i=1}^{\infty} a_i \cos ix,$$

where, by Fourier's theorem,

(15.4) $$\frac{1}{n} = \frac{1}{2\pi} \int_0^{2\pi} \frac{dx}{(C + 2\kappa^2 \cos x)^{\frac{1}{2}}}.$$

This last equation gives a relation between n, C; either may be used as an arbitrary constant.

The value of x in terms of t may be deduced from (15.2), (15.3) after integration. But it is more easily obtained by assuming

$$(15.5) \qquad x = nt + \epsilon + x_1 \equiv nt + \epsilon + \sum_{i=1}^{i=\infty} b_i \sin i(nt+\epsilon),$$

and substituting in (14.1). We have

$$\sin x = \sin(nt+\epsilon)\,\cos x_1 + \cos(nt+\epsilon)\,\sin x_1$$
$$= \sin(nt+\epsilon) \cdot (1 - \tfrac{1}{2}x_1^2 + \cdots)$$
$$+ \cos(nt+\epsilon) \cdot (x_1 - \tfrac{1}{6}x_1^3 + \cdots).$$

On substituting the series (15.5) for x_1 and equating to zero the coefficients of $\sin i(nt+\epsilon)$, we obtain by continued approximation

$$(15.6) \qquad x = nt + \epsilon + \frac{\kappa^2}{n^2}\sin(nt+\epsilon) + \tfrac{1}{8}\frac{\kappa^4}{n^4}\sin 2(nt+\epsilon) + \cdots,$$

in which n, ϵ are the arbitrary constants.

The mean angular velocity of the rod is n. The periodic terms in (15.6) may be regarded as constituting a periodic oscillation about the mean phase $nt+\epsilon$. The physical illustration shows that *the half amplitude of the oscillation is always less than* π, and it evidently diminishes as n increases.

16. (ii) $C < 2\kappa^2$. Here $dx/dt = 0$ and changes sign when $\cos x = -C/2\kappa^2$. If we put $C = -2\kappa^2 \cos c$, x evidently oscillates between the values $x = \pm c$. The equation (15.1) may then be written

$$\left(\frac{dx}{dt}\right)^2 = 4\kappa^2(\sin^2 \tfrac{1}{2}c - \sin^2 \tfrac{1}{2}x),$$

giving

$$(16.1) \qquad t + \text{const.} = \int \frac{dx}{2\kappa(\sin^2 \tfrac{1}{2}c - \sin^2 \tfrac{1}{2}x)^{\frac{1}{2}}}.$$

The transformation

$$\sin \tfrac{1}{2}x = \sin \tfrac{1}{2}c \,\sin \psi,$$

turns (16.1) into

(16.2) $\qquad t + \text{const.} = \displaystyle\int \frac{d\psi}{\kappa(1 - \sin^2 \frac{1}{2}c \cdot \sin^2 \psi)^{\frac{1}{2}}}.$

The integrand can be expanded by the binomial theorem and expressed in a Fourier series with argument ψ. If we put

(16.3)
$$\frac{1}{p} = \frac{1}{\pi} \int_0^\pi \frac{d\psi}{\kappa(1 - \sin^2 \frac{1}{2}c \sin^2 \psi)^{\frac{1}{2}}}$$
$$= \frac{1}{\kappa}\left\{1 + \left(\frac{1}{2}\right)^2 \sin^2 \frac{1}{2}c + \left(\frac{1 \cdot 3}{2 \cdot 4}\right)^2 \sin^4 \frac{1}{2}c + \cdots\right\},$$

we have the period $2\pi/p$ of a complete oscillation expressed in terms of the semi-amplitude c of x. The frequency p diminishes as c increases. It has a lower limit κ corresponding to $c = 0$, but this limit is never reached since $c = 0$ corresponds to equilibrium and we are supposing that the pendulum is in motion.

When c is small we have approximately

(16.4) $\qquad\qquad p = \kappa(1 - \frac{1}{16}c^2 + \cdots).$

To obtain x in terms of t when c is small, it is simplest to expand (15.1) in the form

(16.5) $\qquad\qquad \dfrac{d^2x}{dt^2} + \kappa^2(x - \frac{1}{6}x^3 + \cdots) = 0,$

and to assume as the solution

$$x = \sum_{i=1}^{i=\infty} c_i \sin i(pt + \epsilon).$$

This is substituted in (16.5) and the powers of the series are expressed as sines of multiples of $pt + \epsilon$; evidently only odd multiples will be present. The coefficients of the various sines are then equated to zero and the resulting equations solved by approximation. The coefficient of $\sin(pt + \epsilon)$ equated to zero gives

(16.6) $p = \kappa\left(1 - \frac{c_1^2}{16} + \cdots\right)$,

and c_1 is arbitrary. The solution is found to be

(16.7) $x = c_1 \sin(pt+\epsilon) + \frac{c_1^3}{192} \sin 3(pt+\epsilon) + \cdots$.

The maximum value c of x is given by $pt+\epsilon = 90°$. Hence $c = c_1 - c_1^3/192 + \cdots$.

In the applications to resonance problems it will usually be found sufficient to confine (16.7) to its first term. The only part of the higher approximation which we shall need is the second term of (16.6) and in this we can put $c = c_1$. The presence of this second term is fundamental for the development of the resonance phenomena. The maximum value of c is evidently less than π.

It is still possible to use ϵ and the frequency p as the arbitrary constants instead of ϵ, c. But it is inconvenient, for c is then defined approximately by the expression $4(1-p/\kappa)^{\frac{1}{2}}$ which gives complicated results when derivatives with respect to p are needed.

17. (iii). $C = 2\kappa^2$. This is the limiting case separating cases (i), (ii). Here

$$\left(\frac{dx}{dt}\right)^2 = 4\kappa^2 \cos^2 \tfrac{1}{2}x,$$

the complete solution, containing one arbitrary constant ϵ only, being

$$\kappa t + \epsilon = \log \tan \tfrac{1}{4}(x+\pi),$$

or

(17.1) $x + \pi = 4 \tan^{-1} \exp.(\kappa t + \epsilon)$.

In this case dx/dt oscillates between $\pm 2\kappa$. We have $x \to \pi$ when $t \to \infty$ and $x \to -\pi$ when $t \to -\infty$. At both limiting positions $dx/dt = 0$, $d^2x/dt^2 = 0$. If we form the higher derivatives of x in succession from (15.1), we see that they will all be zero at these limiting places. From the analytical point of view these are singularities of (17.1) and no expansions in powers of t about these points exist. They are evidently the singular points of (15.6) as $n \to 0$ and of (16.7) as $c \to \pi$.

Hence infinitesimal changes in the initial values of x, dx/dt will give different types of motion according to the nature of these changes. If they are within the probable errors of their determination by observation, we must therefore regard the future motion as indeterminate or non-calculable.

18. *Summary.* In the applications of the solutions of (15.1) to resonance problems, emphasis on the following results will be needed.

There are two principal types of solution in one of which the mean value of dx/dt is always positive or always negative but not zero, and in the other of which the mean value of dx/dt is zero.

When the mean value of dx/dt is zero there is a range of solutions in which x oscillates about the value zero, the range being characterized by the half amplitude c which can have any value between 0, π.

The constant ϵ merely gives the origin of reckoning of t. If we suppose this to be settled and then attempt to classify the solutions according to the mean value of dx/dt we find a single solution for each non-zero value and an infinite number for the zero value.

A discontinuity separates the zero from the non-zero values, such that we cannot pass from one set of solutions to the other set by mere changes in the constants.

When the mean value of dx/dt is not zero, the solution has the form

$$(18.1) \qquad x = nt + \epsilon + \frac{\kappa^2}{n^2} \sin(nt + \epsilon) + \tfrac{1}{8} \frac{\kappa^4}{n^4} \sin 2(nt + \epsilon) + \cdots$$

where n, ϵ are the arbitrary constants.

When the mean value of dx/dt is zero, it takes the form

(18.2) $x = c \sin(pt + \epsilon) + \dfrac{c^3}{192} \sin 3(pt + \epsilon) + \cdots;$

with

(18.3) $p = \kappa\left(1 - \dfrac{c^2}{16} + \cdots\right),$

and with c, ϵ as the arbitrary constants.

In (18.1), the development proceeds according to powers of κ^2. In (18.2), it depends on the first power of κ in a quite different manner.

The maximum oscillation of dx/dt occurs at the limit between resonance and non-resonance and its amplitude is 2κ.

METHOD OF APPROXIMATION

19. In the majority of physical problems, defined by differential equations, the only practicable method for obtaining a solution is some process of approximation. On account of either mathematical or physical reasoning, we suppose that some portion of the equation or of its solution may be neglected as a first step, when by this neglect we are able to deduce a result by known methods. Various devices are then available for correcting the result.

One method frequently adopted is a process of continued approximation which will be illustrated by a simple example which has been chosen to show how the method may sometimes fail.

Consider the equation

$$(19.1) \qquad \frac{d^2y}{dt^2} = m \, \sin(y - n't),$$

where m is supposed to be small. If m be neglected, the solution is $y = nt + \epsilon$, where n, ϵ are arbitrary constants. The ordinary meaning attached to this result is that it represents approximately the result we desire to obtain.

The usual procedure is the substitution of this approximate value of y in the right-hand member of (19.1), as giving an approximate value of this term. If we do so and solve again, we obtain

$$(19.2) \qquad y = nt + \epsilon - \frac{m}{(n-n')^2} \sin\{(n-n')t + \epsilon\}.$$

For a further approximation we substitute the more accurate result (19.2) in the right-hand member of (19.1) and solve again. Evidently the process may be continued indefinitely.

It is not difficult to show that the mathematical implication of the process is the possibility of development of the solution in powers of m. If, however, $n-n'$ is very small, the process will evidently not be convergent. This example was chosen because in this case the substitution $y = n't + x + \pi$, transforms the equation to

$$(19.3) \qquad \frac{d^2x}{dt^2} + m \sin x = 0,$$

which is the same as the pendulum equation previously treated if we put $m = \kappa^2$. (If m be negative, the substitution $y = n't + x$ should be used.)

The result shows that when x is an oscillating quantity, y oscillates about $n't + \epsilon$, that is, we must assume $n = n'$. In this case the solution depends on κ or on $m^{\frac{1}{2}}$ and is not developable in positive powers of m. This example illustrates the manner in which certain approximation processes fail under resonance conditions.

THE GENERALIZED EQUATION

20. The discussion which has been carried out for the simple pendulum may be extended to the equation

(20.1)
$$\frac{d^2x}{dt^2} + f'(x) = 0.$$

This may be visualized as the equation of motion of a particle in a smooth closed tube in a vertical plane, the tube having a given shape. Since the discussion of the solutions follows the same lines as that of the simple pendulum, the results will be stated briefly without detailed development.

If $f'(x) = (d/dx) f(x)$, we have the integral

(20.2)
$$\left(\frac{dx}{dt}\right)^2 = C - 2f(x),$$

where C is an arbitrary constant.

We are interested chiefly in those types for which $f'(x)$ vanishes with x. If, further, $f(x)$ has an upper finite limit, there will be at least two types of solution according as C is greater or less than the maximum value of $2f(x)$. In the former case, dx/dt never vanishes and it has a mean value about which it oscillates. In the latter case, it vanishes for at least two values of x; if $f(x)$ be an even function of x these values are equal and opposite, say $x = \pm c$, and then $C = 2f(c) = 2f(-c)$.

In the former case the solution has the form

(20.3)
$$x = nt + \epsilon + \Sigma a_i \sin i(nt + \epsilon),$$

where n, ϵ are arbitrary constants and the coefficients a_i are functions of n.

In the latter case, the solution is

(20.4)
$$x = \Sigma c_i \sin i(pt + \epsilon),$$

where p and the c_i are functions of C.

II

SOLUTIONS OF $\dfrac{d^2x}{dt^2} + f'(x) = m\phi'(x, t)$.

21. This is the type of equation which arises in many resonance problems. In general $m\phi'$ will be small compared with f', and solutions can be obtained when $m = 0$. A process for approximating to the solution when m is not 0 has been described and illustrated in section 19; it was also pointed out that under certain circumstances this process fails. As a matter of fact it fails in those cases which are particularly under investigation here, so that some other method is required.

The method which will be developed in detail below is known in text-books on differential equations as that of the Variation of Arbitrary Constants—a term which conceals its essential characteristics and not infrequently leads to erroneous interpretations. Fundamentally, it consists of a change of variables which is carried out in such a manner as to have the following properties.

(i) Two new variables replace the variable x.

(ii) One relation between x and the new variables is indicated by the solution of a differential equation which can be solved completely. In the present case it will be the equation given above with $m = 0$.

(iii) The replacement of a single variable by two others needs a second relation in order that the change may be definite; this second relation is furnished by the condition that the two differential equations to be

21

satisfied by the new variables shall each be of the first order.

Essentially the method is the same as those known under the names of Jacobi and Hamilton. The canonical forms obtained by the latter are not, however, always useful for the complete solution of a dynamical problem without considerable changes. For this and other reasons, the method will be developed *ab initio*, first with a simple example, and then for the equation which constitutes the heading of this section. Finally a method for the solution of the new equations will be developed.

22. The example referred to at the end of the preceding paragraph is the solution of

$$(22.1) \qquad \frac{d^2x}{dt^2} + \kappa^2 x = m \sin t.$$

The solution of this equation when $m = 0$, that is, of

$$(22.2) \qquad \frac{d^2x}{dt^2} + \kappa^2 x = 0,$$

may be written

$$(22.3) \qquad x = A \cos \kappa t + B \sin \kappa t,$$

where A, B are arbitrary constants.

This suggests a change of variables from x to u, v in which

$$(22.4) \qquad x = u \cos \kappa t + v \sin \kappa t.$$

Since there are two new variables we need a relation connecting them. Let this relation be

$$(22.5) \qquad \frac{du}{dt} \cos \kappa t + \frac{dv}{dt} \sin \kappa t = 0.$$

The variable x will now be replaced by u, v in equation (22.1). Differentiating (22.4) we have

$$(22.6) \qquad \frac{dx}{dt} = \frac{du}{dt} \cos \kappa t + \frac{dv}{dt} \sin \kappa t - \kappa u \sin \kappa t + \kappa v \cos \kappa t$$

$$= -\kappa u \sin \kappa t + \kappa v \cos \kappa t,$$

on account of (22.5).

The relation (22.6) is usually expressed in the form

$$(22.7) \qquad \frac{dx}{dt} = \frac{\partial x}{\partial t},$$

which evidently means that the derivative of x with respect to t has the same form whether we treat u, v as constants or variables.

Differentiating (22.6), we obtain

$$\frac{d^2x}{dt^2} = -\kappa \frac{du}{dt} \sin \kappa t + \frac{dv}{dt} \cos \kappa t - \kappa^2 (u \cos \kappa t + v \sin \kappa t),$$

or, on account of (22.4),

$$(22.8) \qquad \frac{d^2x}{dt^2} + \kappa^2 x = -\kappa \frac{du}{dt} \sin \kappa t + \kappa \frac{dv}{dt} \cos \kappa t.$$

The substitution of this for the left-hand member of (22.1) gives

$$(22.9) \qquad -\kappa \frac{du}{dt} \sin \kappa t + \kappa \frac{dv}{dt} \cos \kappa t = m \sin t.$$

In the place of the equation (22.1), we have the two simultaneous equations (22.5), (22.9), each of the first order. The values of du/dt and dv/dt are easily found from them. They are

$$(22.10) \qquad \frac{du}{dt} = -\frac{m}{\kappa} \sin t \sin \kappa t, \qquad \frac{dv}{dt} = \frac{m}{\kappa} \sin t \cos \kappa t.$$

These are the required equations. It is evident that the change of variables given by (22.4) is equivalent to the assumption that the arbitraries A, B are variable and that (22.5) has the effect of preventing the occurrence of d^2u/dt^2, d^2v/dt^2.

In this simple example, the equations (22.10) are easily integrated if the products are expressed as sums of sines and cosines, we find

$$u = \text{const.} - \frac{m}{2\kappa} \left\{ \frac{\sin(t-\kappa t)}{1-\kappa} - \frac{\sin(t+\kappa t)}{1+\kappa} \right\},$$

$$v = \text{const.} - \frac{m}{2\kappa} \left\{ \frac{\cos(t-\kappa t)}{1-\kappa} + \frac{\cos(t+\kappa t)}{1+\kappa} \right\}.$$

The value of x is obtained by substituting these in (22.4).
If the constants in u, v be denoted by A_o, B_o we obtain

$$x = A_o \cos \kappa t + B_o \sin \kappa t + \frac{m \sin t}{\kappa^2 - 1},$$

a result which may be tested by direct substitution in (22.1).

In this case, the right-hand member of (22.1) does not
contain x. It is evident, however, that the process of changing the variables will be the same as far as (22.10) whatever
the right-hand member may be. In fact it is only when we
arrive at the point where the equations corresponding to
(22.10) have to be integrated, that further devices become
necessary.

In the general problem the partial derivatives $\partial x/\partial t$,
$\partial^2 x/\partial t^2$ will be used. These are to be formed on the assumption that x is expressed as a function of u, v, t. Thus (22.4)
gives

$$\frac{\partial^2 x}{\partial t^2} + \kappa^2 x = 0.$$

23. *Transformation of the equation*

(23.1) $$\frac{d^2 x}{dt^2} + f'(x) = m\phi'(x, t),$$

by the use of the solution of

(23.2) $$\frac{d^2 x}{dt^2} + f'(x) = 0.$$

The solution of (23.2) will contain two arbitrary constants
and is supposed to have been obtained. Let us denote this
solution by

(23.3) $$x = \psi(c, \epsilon, t).$$

Another aspect of the meaning of (23.3), indicated by the
last paragraph of section 22, will be useful. Suppose we
regard (23.3) as defining x in terms of those variables c, ϵ, t
and that we form $\partial^2 x/\partial t^2$, or, what is the same thing, $\partial^2 \psi/\partial t^2$,

which means that t is alone varied in forming the partial derivatives. If this and the value of x be substituted in

(23.4) $$\frac{\partial^2 x}{\partial t^2} + f'(x),$$

the "solution" means that c, ϵ will disappear whatever their meaning and that the expression (23.4) will reduce to zero.

The equation for the transformation from x to the new variables c, ϵ is (23.3). Differentiating it with respect to t we obtain

$$\frac{dx}{dt} = \frac{\partial \psi}{\partial c} \frac{dc}{dt} + \frac{\partial \psi}{\partial \epsilon} \frac{d\epsilon}{dt} + \frac{\partial \psi}{\partial t},$$

or, as it is usually written,

(23.5) $$\frac{dx}{dt} = \frac{\partial x}{\partial c} \frac{dc}{dt} + \frac{\partial x}{\partial \epsilon} \frac{d\epsilon}{dt} + \frac{\partial x}{\partial t}.$$

The additional relation connecting c, ϵ, t, needed to define the change of variables, will be taken to be

(23.6) $$\frac{\partial x}{\partial c} \frac{dc}{dt} + \frac{\partial x}{\partial \epsilon} \frac{d\epsilon}{dt} = 0,$$

so that, in virtue of (23.5),

(23.7) $$\frac{dx}{dt} = \frac{\partial x}{\partial t}.$$

Differentiate (23.7), remembering that $\partial x / \partial t$ is a function of c, ϵ, t. We obtain

(23.8) $$\frac{d^2 x}{dt^2} = \frac{\partial^2 x}{\partial c \partial t} \cdot \frac{dc}{dt} + \frac{\partial^2 x}{\partial \epsilon \partial t} \cdot \frac{d\epsilon}{dt} + \frac{\partial^2 x}{\partial t^2}.$$

Substituting the result in (23.1), and making use of the fact that the expression (23.4) is zero, we obtain

(23.9) $$\frac{\partial^2 x}{\partial c \partial t} \cdot \frac{dc}{dt} + \frac{\partial^2 x}{\partial \epsilon \partial t} \cdot \frac{d\epsilon}{dt} = m\phi'(x, t).$$

Equations (23.6), (23.9) may be regarded as two simultaneous equations for finding dc/dt, $d\epsilon/dt$. On solving them as such we obtain

$$(23.10) \quad \frac{dc}{dt} = \frac{m}{K}\frac{\partial x}{\partial \epsilon}\phi'(x,t), \quad \frac{d\epsilon}{dt} = -\frac{m}{K}\frac{\partial x}{\partial c}\phi'(x,t),$$

where

$$(23.11) \qquad K \equiv \frac{\partial^2 x}{\partial c\partial t}\cdot\frac{\partial x}{\partial \epsilon} - \frac{\partial^2 x}{\partial \epsilon\partial t}\cdot\frac{\partial x}{\partial c}.$$

When the right-hand members of (23.10) have been expressed in terms of c, ϵ, t by the use of (23.3), these equations become a pair of differential equations of the first order for finding c, ϵ in terms t. The values thus found are substituted in (23.3) and give the value of x.

The divisor K does not contain t explicitly. This important property is deduced from the fact that (23.4) is zero whatever c, ϵ, may be and therefore that its partial derivatives with respect to c, ϵ are also zero. Hence

$$\frac{\partial^3 x}{\partial c\partial t^2} + \frac{\partial f'}{\partial x}\cdot\frac{\partial x}{\partial c} = 0, \qquad \frac{\partial^3 x}{\partial \epsilon\partial t^2} + \frac{\partial f'}{\partial x}\cdot\frac{\partial x}{\partial \epsilon} = 0.$$

The elimination of $\partial f'/\partial x$ from these gives

$$\frac{\partial^3 x}{\partial c\partial t^2}\cdot\frac{\partial x}{\partial \epsilon} - \frac{\partial^3 x}{\partial \epsilon\partial t^2}\cdot\frac{\partial x}{\partial c} = 0,$$

and this may be written

$$\frac{\partial}{\partial t}\left(\frac{\partial^2 x}{\partial c\partial t}\cdot\frac{\partial x}{\partial \epsilon} - \frac{\partial^2 x}{\partial \epsilon\partial t}\cdot\frac{\partial x}{\partial c}\right) = 0,$$

or $\partial K/\partial t = 0$, which proves the statement.

It is evident from (23.10) that c, ϵ become constants when $m = 0$, and therefore that the solution of (23.10) gives their variations due to the presence of the term $m\phi'$; this point of view, as stated before, is responsible for the term "variation of arbitrary constants."

It is to be noticed also that the relations

$$x = \psi(c, \epsilon, t), \quad \frac{dx}{dt} = \frac{\partial \psi}{\partial t} \equiv \frac{\partial x}{\partial t},$$

may be interpreted as meaning that not only x, but also dx/dt, have the same *form* when expressed as functions of c, ϵ, t, and whether c, ϵ be variable or constant. This is not true of d^2x/dt^2.

24. *Second change of variables.* If the solution of $d^2x/dt^2 + f'(x) = 0$ has either of the forms given in section 20 difficulties may arise when the variable values of c, ϵ are substituted in the expression for x. For example, if in (20.3), we choose the arbitraries n, ϵ as our new variables, it may turn out that n is a periodic function of t and we thus apparently have terms of the form t multiplied by a periodic function of t, which we know will disappear from the final expression. The presence of such terms can be avoided by a change of variables somewhat different from that used in section 23.

Either of the solutions (20.3) or (20.4) may be expressed in the form

$x = $ function of $c, l; l = nt + \epsilon; n = $ function of $c,$
where c, ϵ represent the adopted arbitrary constants.

Since t is present in x only through its presence in l, we have

$$\frac{\partial x}{\partial t} = \frac{\partial x}{\partial l} \cdot \frac{\partial l}{\partial t} = n \frac{\partial x}{\partial l}, \quad \frac{\partial^2 x}{\partial t^2} = n^2 \frac{\partial^2 x}{\partial l^2}.$$

Thus the equation $\partial^2x/\partial t^2 + f'(x) = 0$ can be written

(24.1) $$n^2 \frac{\partial^2 x}{\partial l^2} + f'(x) = 0,$$

where x is now a function of c, l and not of t; and n is the function of c previously defined.

Changes of Variables

The new variables will now be c, l and the additional relation needed will be chosen to be

(24.2)
$$\frac{dx}{dt} = n\frac{\partial x}{\partial l}.$$

Since

(24.3)
$$\frac{dx}{dt} = \frac{\partial x}{\partial c} \cdot \frac{\partial c}{dt} + \frac{\partial x}{\partial l} \cdot \frac{dl}{dt},$$

we have

(24.4)
$$\frac{\partial x}{\partial c} \cdot \frac{dc}{dt} + \frac{\partial x}{\partial l}\left(\frac{dl}{dt} - n\right) = 0.$$

Also since n is a function of c only and $\partial x/\partial l$ a function of c, l, we have, from the differentiation of (24.2),

$$\frac{d^2 x}{dt^2} = \frac{\partial}{\partial c}\left(n\frac{\partial x}{\partial l}\right) \cdot \frac{dc}{dt} + n\frac{\partial^2 x}{\partial l^2} \cdot \frac{dl}{dt}.$$

Substituting this in (23.1) and making use of (24.1), we obtain

(24.5)
$$\frac{\partial}{\partial c}\left(n\frac{\partial x}{\partial l}\right) \cdot \frac{dc}{dt} + n\frac{\partial^2 x}{\partial l^2}\left(\frac{dl}{dt} - n\right) = m\phi'(x, t).$$

The equations (24.4), (24.5) may be regarded as simultaneous for the determination of dc/dt, $(dl/dt) - n$. Their solution gives

(24.6)
$$\frac{dc}{dt} = \frac{m}{K}\frac{\partial x}{\partial l}\phi', \quad \frac{dl}{dt} = n - \frac{m}{K}\frac{\partial x}{\partial c}\phi',$$

where

(24.7)
$$K \equiv \frac{\partial x}{\partial l} \cdot \frac{\partial}{\partial c}\left(n\frac{\partial x}{\partial l}\right) - n\frac{\partial x}{\partial c} \cdot \frac{\partial^2 x}{\partial l^2}.$$

The substitution for x of its value in terms of c, l and of n in terms of c, gives us two equations of the first order for the determination of l, c in terms of t.

The proof that K is a function of c only can be obtained as in section 23. Another proof is as follows:—

On account of (24.1) we can write (24.7) in the forms

$$Kn = n\frac{\partial x}{\partial l} \cdot \frac{\partial}{\partial c}\left(n\,\frac{\partial x}{\partial l}\right) + \frac{\partial x}{\partial c}f'(x)$$

$$= \frac{\partial}{\partial c}\left\{\frac{1}{2}\left(n\,\frac{\partial x}{\partial l}\right)^2 + f(x)\right\}$$

$$= \frac{\partial}{\partial c}\left\{\frac{1}{2}\left(\frac{\partial x}{\partial t}\right)^2 + f(x)\right\},$$

by (24.2). But the integral of $d^2x/dt^2 + f'(x) = 0$ is

(24.8)
$$\frac{1}{2}\left(\frac{dx}{dt}\right)^2 + f(x) = \frac{1}{2}C,$$

where C is evidently a function of c only. Thus

(24.9)
$$K = \frac{1}{2n}\,\frac{\partial C}{\partial c}.$$

25. We have supposed that ϕ' was a function of x, t only. There is nothing in the transformation to prevent it from being a function of dx/dt also, since on account of the relation $dx/dt = n\partial x/\partial l$, this derivative can be expressed as a function of l, c. Thus the transformation can be used when frictional forces depending on the velocity are present.

When, however, ϕ' is a function of x, t only we can write

$$\phi' = \frac{\partial \phi}{\partial x},$$

and the equations (24.6) can then be written

(25.1)
$$\frac{dc}{dt} = \frac{m}{K}\,\frac{\partial \phi}{\partial l}, \quad \frac{dl}{dt} = n - \frac{m}{K}\,\frac{\partial \phi}{\partial c},$$

a form which saves much laborious calculation, since ϕ only has to be expressed in terms of c, l, and K is usually a quite simple function of c. It is recalled that l occurs in ϕ only through its presence in the expression which gives x as a function of c, l.

Although the canonical form of these equations will not be used here, it may be deduced immediately by defining new variables c_1, B, with the equations,

$$dc_1 = K\,dc, \qquad dB = -n\,dc_1 = -nK\,dc.$$

Since B is independent of l, the equations (25.1) can then be written

$$\frac{dc_1}{at} = \frac{\partial}{\partial l}(m\phi + B), \quad \frac{dl}{dt} = -\frac{\partial}{\partial c}(m\phi + B).$$

Equation (24.9) shows that $B = -\frac{1}{2}C$ and that $dc_1 = dC/2n$.

26. *Approximate Solution of the Equations for c, l.* When the equations (25.1) have been formed and a solution is needed, it is important to remember that n is a function of c, and therefore that if an approximation process be adopted, the equation for n must be solved before that for l.

We have seen, however, that in resonance problems it may not be possible to follow the usual processes because developments in powers of m fail to be convergent. The following plan will often be found effective in such cases.

Differentiate the equation for l and substitute in the result the expressions for dc/dt, dl/dt, remembering that ϕ may be a function of t as well as of c, l. We obtain

$$\frac{d^2 l}{dt^2} = \frac{\partial}{\partial c}\left(n - \frac{m}{K}\frac{\partial\phi}{\partial c}\right)\cdot\frac{dc}{dt} - \frac{m}{K}\frac{\partial^2\phi}{\partial l\partial c}\cdot\frac{dl}{dt} - \frac{m}{K}\frac{\partial^2\phi}{\partial c\partial t}$$

$$(26.1) \qquad = \frac{m}{K}\left(\frac{\partial n}{\partial c}\cdot\frac{\partial\phi}{\partial l} - n\frac{\partial^2\phi}{\partial l\partial c} - \frac{\partial^2\phi}{\partial c\partial t}\right)$$

$$+ \frac{m^2}{K^2}\left\{\frac{\partial^2\phi}{\partial l\partial c}\cdot\frac{\partial\phi}{\partial c} - K\frac{\partial}{\partial c}\left(\frac{1}{K}\frac{\partial\phi}{\partial c}\right)\cdot\frac{\partial\phi}{\partial l}\right\}.$$

In a first approximation, the terms factored by m^2 will be neglected. The equation for d^2c/dt^2 might also be formed but it will not be needed.

The problems to be considered are those in which, after x has been given its value in terms of c, l, we have a term in ϕ of the form

(26.2) $$\phi = -a_i \cos(il - n't - \epsilon'),$$

where a_i is a function of c only; n', ϵ' are given constants, and i is an integer. We then have

$$n \frac{\partial^2 \phi}{\partial l \partial c} + \frac{\partial^2 \phi}{\partial t \partial c} = \left(n - \frac{n'}{i}\right) \frac{\partial^2 \phi}{\partial l \partial c}$$

$$= (in - n') \frac{da_i}{dc} \sin (il - n't - \epsilon').$$

This result enables us to write (26.1) (without its last line) in the form

(26.3) $$\frac{d^2 l}{dt^2} + \frac{m}{K} \left\{ -ia_i \frac{dn}{dc} + (in - n') \frac{da_i}{dc} \right\} \sin(il - n't - \epsilon') = 0,$$

or

(26.3a) $$\frac{d^2 l}{dt^2} + \frac{m}{K} (in - n')^2 \frac{\partial}{\partial c}\left(\frac{a_i}{in - n'}\right) \sin(il - n't - \epsilon') = 0.$$

Finally, if we put

$$il - n't - \epsilon' = l_i \text{ or } l_i + \pi$$

according as

(26.4) $$\frac{m}{K} \frac{\partial}{\partial c}\left(\frac{a_i}{in - n'}\right) > 0 \text{ or } < 0,$$

the equation will take the form

(26.5) $$\frac{d^2 l_i}{dt^2} + p^2 \sin l_i = 0,$$

where

(26.6) $$p^2 \equiv \left| i(in - n') \frac{m}{K} \frac{da_i}{dc} - i^2 \frac{dn}{dc} \cdot \frac{m}{K} a_i \right|,$$

or

(26.6a) $$p^2 \equiv \left| i(in - n')^2 \frac{m}{K} \frac{\partial}{\partial c}\left(\frac{a_i}{in - n'}\right) \right|.$$

It will be shown below that a first approximation to the solution of (26.5) may be obtained by putting c equal to a constant c_0; this assumption makes n, K, a_i, constants. Thus p^2 is constant *and* (5) *becomes the same as the characteristic equation for the motion of a simple pendulum.*

27. We can therefore make use of the discussion given in sections 14–18.

If l_i makes complete revolutions with a mean angular velocity $in_0 - n'$, that is, if

(27.1) $il = i(n_0t + \epsilon_0) + \delta l_i ,$

in which the arbitrary constant n_0 has been so chosen that $in_0 - n'$ is not 0, and δl_i is an oscillating function, we have the analogue to the case in which the pendulum is making complete revolutions. When the departure of l_i from its mean value is small, the integration of the equation gives approximately (18.1),

(27.2) $\delta l_i = \left\{ \dfrac{mi}{K} \dfrac{\partial}{\partial c} \left(\dfrac{a_i}{in - n'} \right) \right\}_0 \sin(in_0t + i\epsilon_0 - n't - \epsilon').$

This is the non-resonance case. The suffix zero denotes that c has been put equal to c_0 in the inclosed expression.

The resonance case is that in which l_i is an oscillating angle; it corresponds to the case in which the pendulum is oscillating. Accordingly, we must have

(27.3) $in_0 - n' = 0, \; i\epsilon_0 = \epsilon'$ or $\epsilon' + \pi,$

since we saw that the pendulum must oscillate about one of the values 0 or π of the variable.

With the use of the form (26.6) and the insertion of $in_0 = n'$, we have

(27.4) $p^2 = \left| i^2 \dfrac{m}{K} \dfrac{dn}{dc} a_i \right|_0 .$

When the amplitude of the oscillation is small, we have (18.2)

(27.5) $$il - n't - \epsilon' \equiv l_i = \lambda \sin(pt + \lambda_0),$$

where λ, λ_0 are arbitrary constants. In certain physical problems this oscillation is called a "libration."

It is to be noticed that the original arbitrary constants c_0, ϵ_0 have become definite, for c_0 is defined by $in_0 - n' = 0$, where n_0 is a known function of c_0, and $\epsilon_0 = \epsilon'$ or $\epsilon' + \pi$. They are replaced as arbitrary constants by λ, λ_0.

Just as in the motion of the pendulum an essential singularity separates the solutions for complete revolution and for oscillation about the vertical, so a similar singularity separates the non-resonance from the resonance case. So long as we confine our work to this first approximation the two problems give analogous results.

28. The solution which gives a first approximation to l has been so carried out that we were able to neglect the variation of c in finding it. This variation has still to be obtained.

Equations (25.1), (26.2) give

(28.1) $$\frac{dc}{dt} = \frac{m}{K} \frac{\partial \phi}{\partial l} = \frac{m}{K} ia_i \sin l_i.$$

Since the right-hand member has the factor m, we shall neglect the variation of c therein. The use of (26.5) then gives

$$\frac{dc}{dt} = - \left(\frac{mia_i}{Kp^2}\right)_0 \frac{d^2 l_i}{dt^2},$$

which, on integration, furnishes

(28.2) $$c = \text{const.} - \left(\frac{mia_i}{Kp^2}\right)_0 \frac{dl_i}{dt}.$$

In this equation one of the values of l_i obtained in section 27 is to be used.

In the non-resonance case, the constant part of dl_i/dt may be supposed to be absorbed into the arbitrary constant.

The oscillating part is given by (27.2) and it has the factor m. But p^2 also has the factor m. Thus the oscillating term in c has the factor m and the earlier assumption that it may be neglected when multiplied by the factor m leads to no contradiction.

In the resonance case, when the libration is small, the substitution of (27.5) in (28.2) gives

$$(28.3) \qquad c = c_0 - \left(\frac{mia_i}{Kp}\right)_0 \lambda \cos (pt+\lambda_0),$$

where c_0 is the constant part of c. The insertion of the value (27.4) gives for the coefficient of the periodic term,

$$(28.4) \qquad \left| \frac{ma_i}{K} \div \frac{dn}{dc} \right|_0^{\frac{1}{2}} \lambda.$$

This expression has the factor $m^{\frac{1}{2}}$ and therefore, provided no other part of the coefficient tends to become large when the adopted value of c_0 is used, it will be small, and again leads to no contradiction in our earlier assumptions.

We recall that c_0 in this case is not arbitrary but is determined by solving the equation $in_0 = n'$, where n_0 is a known function of c_0. The two arbitrary constants needed in the solution have been already shown to be λ, λ_0.

Attention is again directed to the fact that in non-resonance cases, the expansions proceed according to powers of m, but that in resonance cases they start with $m^{\frac{1}{2}}$, when (28.4) does not tend to become large. If this last condition is not satisfied the approximation may not be valid and some other procedure for the solution of the equations will have to be devised. An example of this will be shown below.

29. The variable l is the phase of the periodic motion when $m = 0$, while c is connected with the amplitude of this motion. We have seen that in non-resonance cases, l, c both

receive variations which have m as a factor, so that these variations will in general be small, although they tend to become large as a resonance range is approached.

In resonance, the circumstances are quite different. The oscillation of l has an arbitrary amplitude, and can apparently be of any magnitude. But the analogy with the motion of the pendulum shows that, in general, the oscillation of l will be between $\pm\pi$ in the limiting case between resonance and non-resonance, or at least will be always finite and of this order of magnitude. The rate of change of this oscillation is however always small.

On the other hand, the oscillation of c has in general a small amplitude in both resonance and non-resonance cases and has $m^{\frac{1}{2}}$ or m as a factor. By (28.2), its amplitude of oscillation is a maximum when that of dl_i/dt is a maximum; according to section 18, this latter maximum is $2p_0$. The maximum amplitude of oscillation of c is therefore

$$(29.1) \qquad 2\left|\frac{mia_i}{Kp}\right|_0 = 2\left|\frac{ma_i}{K\dfrac{dn}{dc}}\right|_0^{\frac{1}{2}}.$$

The importance of this result lies in the fact that *there is no tendency for c to become infinite*, as often assumed, at least in the types of resonance indicated by the equations we have used above. Thus there is no need of the damping factor often introduced to avoid this difficulty. In fact, as we shall see below, the presence of a damping factor may in certain cases ensure the passage from non-resonance to resonance, and thus actually produce the phenomena which its introduction was intended to avoid.

III

APPLICATION TO THE PENDULUM WITH AN OSCILLATING SUPPORT

30. The first problem which we shall consider is the motion of an ordinary pendulum which is subjected to a type of disturbance liable to produce resonance effects.

For simplicity, we shall suppose that the pendulum is the ideal simple one of length b. The disturbance is communicated through the point of support S which is supposed to be movable in a horizontal direction only.

Let y be the horizontal distance of S from a fixed point O and x the angle which the pendulum makes with the

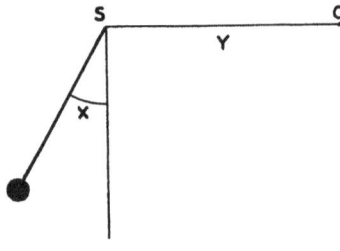

FIG. I

vertical at time t. The equation of motion of the pendulum is then

(30.1)
$$\frac{d^2y}{dt^2} \cos x + b \frac{d^2x}{dt^2} = -g \sin x.$$

Suppose that S is forced to oscillate with a motion defined by

(30.2)
$$y = -b' \sin(n't + \epsilon'),$$

where b', n', ϵ' are given. It is thus supposed to be unaffected by the oscillation of the pendulum.

With this value for y, equation (30.1) becomes

$$b\,\frac{d^2x}{dt^2} + g\sin x = -n'^2b'\sin(n't+\epsilon')\cos x$$

(30.3)
$$= -\,n'^2b'\,\frac{\partial}{\partial x}\Big\{\sin(n't+\epsilon')\sin x\Big\}.$$

If we put

(30.4)　$g/b = \kappa^2$, $n'^2b' = \kappa^2 mb$, $\phi = -\kappa^2\sin(n't+\epsilon')\sin x$,

equation (30.3) may be written

(30.5)
$$\frac{d^2x}{dt^2} + \kappa^2\sin x = m\,\frac{\partial\phi}{\partial x}.$$

31. In order to solve (30.5) according to the method developed above, we first solve the equation when $m=0$. The solution has already been obtained. As we shall suppose that the pendulum is performing oscillations with not very large amplitude, the solution is that given in section 16, namely,

(31.1)
$$\begin{cases} x = c\sin l + \dfrac{c^3}{192}\sin 3l + \cdots \\[2mm] n = \kappa\Big(1 - \dfrac{c^2}{16} + \cdots\Big),\ \dfrac{dx}{dt} = n\,\dfrac{\partial x}{\partial l}, \end{cases}$$

these values satisfying

(31.2)
$$n^2\,\frac{\partial^2 x}{\partial l^2} + \kappa^2\sin x = 0$$

for all values of c, l.

When m is not 0, we choose c, l as the new variables and, according to (25.1), they satisfy the equations

(31.3)
$$\frac{dc}{dt} = \frac{m\partial\phi}{K\partial l}, \quad \frac{dl}{dt} = n - \frac{m\partial\phi}{K\partial c}$$

where K is given by (24.7).

The calculation of K is quite easy if we make use of the fact that it is independent of l. Let us put $l=90°$ after forming the derivatives from (31.1). The first term of (24.7) vanishes and for the second term we have, when $l=90°$,

$$\frac{\partial x}{\partial c} = 1 - \frac{c^2}{64} + \cdots , \quad \frac{\partial^2 x}{\partial l^2} = -c + \frac{3c^3}{64} \cdots ,$$

so that

(31.4) $$K = \kappa c \left(1 - \frac{c^2}{8} + \cdots \right).$$

Next, ϕ has to be expressed in terms of c, l. The simplest way to do this is to calculate $\sin x$ from (31.2) with the help of (31.1). We find

$$\sin x = - \frac{n^2 \partial^2 x}{\kappa^2 \partial l^2} = \frac{n^2}{\kappa^2} \left(c \sin l + \frac{3c^3}{64} \sin 3l + \cdots \right).$$

Hence, from (30.4),

(31.5) $\begin{cases} \phi = -\tfrac{1}{2} n^2 c \, \cos(l - n't - \epsilon') + \tfrac{1}{2} n^2 c \, \cos(l + n't + \epsilon') \\ \quad - \tfrac{3}{128} n^2 c^3 \, \cos(3l - n't - \epsilon') \\ \quad + \tfrac{3}{128} n^2 c^3 \, \cos(3l + n't + \epsilon') + \cdots . \end{cases}$

The substitution of this value of ϕ in (31.3) gives the required equations for finding c, l.

32. Since m is supposed to be small and since we shall neglect m^2, the various terms in ϕ may be regarded as separate disturbances, each producing its own effect: the total effect being the sum of the separate portions in this approximation.[1]

The chief interest centers on the first term because it gives resonance phenomena when $n = n'$. For this term the solu-

---

[1] The method of Delaunay, as used in celestial mechanics, avoids this assumption.

tion in sections 26–29 may be directly used. We have, in fact,

$$l_i = l - n't - \epsilon', \ i = 1, \ a_i = \tfrac{1}{2}n^2c.$$

Also, from (31.1),

$$\frac{dn}{dc} = -\tfrac{1}{8}\kappa c + \cdots.$$

Following the notation of these sections, we have $n_0 = n'$, so that by (31.1)

$$(32.1) \qquad c_0 = 4\left(\frac{\kappa - n'}{\kappa}\right)^{\frac{1}{2}},$$

approximately.

Hence a first approximation shows that an oscillation of the pendulum with frequency n'—the same as that of the disturbing force—is possible provided the arc through which it oscillates is given by (32.1). Evidently, it is necessary that $\kappa > n'$, since this arc must be greater than zero in order that motion shall occur.

This last condition exhibits the necessity for dealing with the non-linear equation for the motion of the pendulum. If we had used the linear form, the frequency n_0 would have been put equal to κ, and there would have been no clue to the value of the amplitude under resonance conditions.

The frequency p of a small oscillation about this resonance configuration is given by (27.4). In the present case it is approximately given by

$$p^2 = \left| \begin{matrix} m & \dfrac{\kappa c}{8} \\ \kappa c & \end{matrix} \cdot \tfrac{1}{2}\kappa^2 c \right|_0 = \frac{m\kappa^2 c_0}{16}.$$

Hence we have, from (32.1),

$$p = \frac{m^{\frac{1}{2}}\kappa}{2}\left(\frac{\kappa - n'}{\kappa}\right)^{\frac{1}{4}}.$$

The small oscillation of l is given by
$$l = n't + \epsilon' + \lambda \sin(pt + \lambda_0),$$
and that of c by (section 28)
$$c = c_0 - \frac{ma_1}{Kp} \lambda \cos(pt + \lambda_0),$$
where c_0 has the value (32.1). This last coefficient is
$$\frac{m^{\frac{1}{2}} \kappa^2 c_0}{\kappa c_0} \cdot \frac{2}{m^{\frac{1}{2}} \kappa} \cdot \left(\frac{\kappa}{\kappa - n'} \right)^{\frac{1}{2}} \lambda = m^{\frac{1}{2}} \left(\frac{\kappa}{\kappa - n'} \right)^{\frac{1}{2}} \lambda.$$
The maximum amplitude of oscillation of c is given by
(29.1). This maximum is found to be
$$2m^{\frac{1}{2}} \frac{n'}{\kappa} \left(\frac{\kappa}{\kappa - n'} \right)^{\frac{1}{2}}.$$

In order that the approximations may be valid, it is evidently necessary that $\kappa - n'$ shall not be too small.

33. When $\kappa = n'$, the method adopted above fails, and in this case it appears to be easier to return to the original equation (30.3). This, with the notation (30.4), can be written

(33.1) $\qquad \dfrac{d^2x}{dt^2} + \kappa^2 \sin x + \kappa^2 m \cos x \sin(\kappa t + \epsilon') = 0.$

When x is small it is possible to expand $\sin x$, $\cos x$ in powers of x. Let us see whether a solution with period $2\pi/\kappa$ can be found; if so, only odd multiples of $\kappa t + \epsilon'$ will be present. Suppose
$$x = A_1 \sin(\kappa t + \epsilon') + A_3 \sin 3(\kappa t + \epsilon') + \cdots,$$
where A_1, A_3, are small. We have
$$\cos x = 1 - \tfrac{1}{2}x^2 + \cdots = 1 - \tfrac{1}{4}A_1^2 + \tfrac{1}{4}A_1^2 \cos 2(\kappa t + \epsilon') + \cdots$$
$$\sin x = x - \tfrac{1}{6}x^3 + \cdots = A_1 \sin(\kappa t + \epsilon') + A_3 \sin 3(\kappa t + \epsilon') + \cdots$$
$$- \tfrac{1}{8}A_1^3 \sin(\kappa t + \epsilon') + \tfrac{1}{24} A_1^3 \sin 3(\kappa t + \epsilon') + \cdots.$$
Inserting these in (33.1) and equating to zero the coefficients of $\sin(\kappa t + \epsilon')$, $\sin 3(\kappa t + \epsilon')$, we obtain
$$- A_1 \kappa^2 + \kappa^2(A_1 - \tfrac{1}{8}A_1^3) + m\kappa^2(1 - \tfrac{3}{8}A_1^2) + \cdots = 0,$$
$$- 9A_3 \kappa^2 + \kappa^2(A_3 + \tfrac{1}{24}A_1^3) + \tfrac{1}{8}m\kappa^2 A_1^2 + \cdots = 0.$$

The first approximation to the solution of these equations gives

$$A_1 = 2m^{1/3}, \qquad A_3 = \tfrac{1}{192} A_1{}^3 = \frac{m}{24}.$$

It is evident that the coefficient of A_i will have $m^{1/3}$ as a factor, that the approximations proceed along powers of $m^{2/3}$, and that the series converge.

Thus a solution which depends on $m^{\frac{1}{3}}$ exists. This, however, is a *particular* solution since it contains no arbitrary constants. Let us call it $x = x_0$, and see whether a solution $x = x_0 + \delta x$ can be found. On inserting this value of x in (33.1) and neglecting powers of δx beyond the first, we obtain

$$\frac{d^2}{dt^2} \delta x + \{\kappa^2 \cos x_0 - \kappa^2 m \sin x_0 \sin(\kappa t + \epsilon')\} \delta x = 0.$$

The principal terms in the coefficient of δx are those deduced by putting

$$\cos x_0 = 1 - \tfrac{1}{2} x_0{}^2 = 1 - \tfrac{1}{4} A_1{}^2 + \tfrac{1}{4} A_1{}^2 \cos 2(\kappa t + \epsilon'),$$

where $A_1 = 2m^{\frac{1}{3}}$. In general, the coefficient of δx will be a series of the form,

$$1 + \sum_0^\infty b_{2i} \cos 2i(\kappa t + c'),$$

where b_{2i} vanishes with m and b_0 is not zero.

Equations of this type are well known. They give oscillatory motion with an arbitrary amplitude and phase. The principal frequency present depends on b_0. Thus small variations from the particular solution appear to be possible.

34. Hence when $\kappa - n'$ is zero, the resonance case gives expansions in powers of $m^{\frac{1}{3}}$, while we saw in the first part, when $\kappa - n'$ was not too small, that the expansions proceeded along powers of $m^{\frac{1}{2}}$. The theory of analytic forms suggests that a singularity may separate these two sets of solutions.

The method of investigation followed in section 33 might also have been used in the earlier case where $\kappa - n'$ was not zero. It is difficult, however, to get a clue with this method to the transition from non-resonance to resonance and, in any case, it is much less adaptable to the more complicated cases presented in other problems.

Another exercise which may be left to the student is the investigation of the resonances arising when one of the relations $3n - n' = 0$, $5n - n' = 0$, $\cdots$, is approximately satisfied. These resonances require the consideration of the terms with arguments $3l - n't - \epsilon'$, $\cdots$, in (31.5). It can be shown that as long as we retain only the lowest power of m present, each can be treated as though the remaining terms did not exist.

IV

APPLICATIONS TO THE MUTUAL INFLUENCE
OF TWO-PENDULUMS

35. Let us consider the problem of two ideal pendulums, the bobs being treated as particles of masses μ, μ', and the rods as weightless and having lengths b, b'. They are hung from points attached to a bar of mass M capable of horizontal motion only: all frictional effects are neglected.

We shall first suppose that the bar is not acted on by any external force. If each body in the system has zero velocity initially, the horizontal component of the momentum is zero; also the total energy remains unchanged throughout the

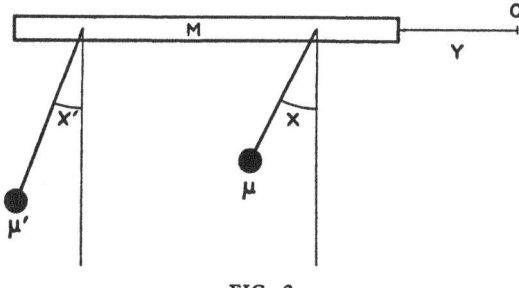

FIG. 2

motion. Let x, x' be the angles which the rods make with the vertical at time t and y the horizontal distance of the center of mass of M from a fixed point O at the same instant. The assumed zero value of the total horizontal momentum gives

(35.1) $My + \mu(y + b \sin x) + \mu'(y + b' \sin x') = \text{const.}$

The forces and accelerations resolved perpendicularly to the rods give, after division by the masses,

$$(35.2) \qquad \frac{d^2y}{dt^2}\cos x + b\frac{d^2x}{dt^2} = -g\sin x,$$

$$(35.3) \qquad \frac{d^2y}{dt^2}\cos x' + b'\frac{d^2x'}{dt^2} = -g\sin x'.$$

The substitution of (35.1) in (35.2) furnishes an equation which may be written

$$(35.4) \quad b\frac{d^2x}{dt^2} + g\sin x - \frac{\mu b}{M+\mu+\mu'}\cos x\frac{d^2}{dt^2}(\sin x)$$
$$= \frac{\mu'b'}{M+\mu+\mu'}\cos x\frac{d^2}{dt^2}(\sin x').$$

If the right-hand member of (35.4) be neglected, the equation becomes integrable on multiplication by dx/dt. Since $\cos x \cdot dx/dt = d(\sin x)/dt$, this integral is

$$\tfrac{1}{2}b\left(\frac{dx}{dt}\right)^2 - g\cos x - \tfrac{1}{2}\frac{\mu b}{M+\mu+\mu'}\left(\cos x\frac{dx}{dt}\right)^2 = \text{const.} = \tfrac{1}{2}C,$$

or

$$\left(\frac{ax}{dt}\right)^2 = (C+2g\cos x)\div b\left(1 - \frac{\mu}{M+\mu+\mu'}\cos^2 x\right) \equiv -2f(x).$$

On differentiating this result we obtain

$$(35.5) \qquad \frac{d^2x}{dt^2} + f'(x) = 0,$$

and $f(x)$ may be written in the form

$$(35.6) \qquad f(x) = -\frac{M+\mu+\mu'}{2b}\cdot\frac{C+2g\cos x}{M+\mu'+\mu\sin^2 x}.$$

Since $f(x)$ lies between finite limits and has no singularity, equation (35.6) is of the type considered in section 20, and the analysis of that and of the succeeding sections may be applied. But the calculations may be much simplified by making approximations which can be shown to be sufficient to investigate the resonance phenomena when the latter are present.

36. The solutions we seek are those in which x, x' are oscillating through angles which are never very great, and in which μ/M, μ'/M are small. The approximations will be made on the assumption that powers and products of x^2, x'^2, μ/M, μ'/M beyond the first can be neglected in $f(x)$: this involves the neglect of products of x^3, x'^3 by μ/M, μ'/M in $f'(x)$.

With these limitations $f(x)$ reduces to

$$f(x) = \text{const.} - \frac{M+\mu+\mu'}{b(M+\mu')} g \cos x ,$$

so that (35.5) becomes the equation of motion of a simple pendulum of length $b(M+\mu') \div (M+\mu+\mu')$. Further, in the right-hand member of (35.4) we can put $\cos x = 1$, $\sin x' = x'$. If then the notation

$$\kappa^2 = \frac{g}{b} \frac{M+\mu+\mu'}{M+\mu'} , \quad \kappa'^2 = \frac{g}{b} \frac{M+\mu+\mu'}{M+\mu} ,$$

$$m = \frac{\mu'}{M+\mu} , \quad m' = \frac{\mu}{M+\mu'} ,$$

be adopted, (35.4) and the similar equation for x' may be written

(36.1)
$$\frac{d^2x}{dt^2} + \kappa^2 \sin x = m \frac{\kappa^2}{\kappa'^2} \frac{d^2x'}{dt^2} ,$$

(36.2)
$$\frac{d^2x'}{dt^2} + \kappa'^2 \sin x' = m' \frac{\kappa'^2}{\kappa^2} \frac{d^2x}{dt^2} .$$

Finally, the neglect of the product mm' enables us to substitute for d^2x'/dt^2 in (36.1), its value derived from (36.2) when the right-hand member of the latter is neglected: equation (36.2) may be similarly treated, and we may put $\sin x = x$, $\sin x' = x'$ in the right-hand members. We then obtain

(36.3)
$$\frac{d^2x}{dt^2} + \kappa^2 \sin x = - m \kappa^2 x' = - m\kappa^2 \frac{\partial}{\partial x} (xx'),$$

(36.4) $\quad \dfrac{d^2x'}{dt^2} +\kappa'^2 \sin x' = -m'\kappa'^2 x = -m'\kappa'^2 \dfrac{\partial}{\partial x'}(xx').$

Each of these equations now has the form ready for the application of the methods developed above.

37. When $m=m'=0$, each of the equations (36.3), (36.4) reduces to that for the motion of a simple pendulum. The variable x is then changed to the variables c, l as in the previous example, and x' to similar variables c', l'. Also, as in the previous example, it is sufficient for a first approximation to confine the solution (36.3) when $m=0$ to

(37.1) $\qquad x=c \sin l, \quad n=\kappa\left(1-\dfrac{c^2}{16}\right), \quad K=\kappa c;$

and the solution of (36.4) when $m'=0$ to

(37.2) $\quad x'=c' \sin l', \quad n'=k'\left(1-\dfrac{c'^2}{16}\right), \quad K'=\kappa'c'.$

If, in the general formulae, we replace m by $m\kappa^2$, and by $m'\kappa'^2$ for the respective equations, we shall then have

$$\phi = -xx' = -cc' \sin l \sin l'$$
(37.3)
$$= -\tfrac{1}{2}cc' \cos(l-l') + \tfrac{1}{2}cc' \cos(l+l').$$

The resonance case corresponds to that in which dl/dt $-dl'/dt$ is nearly zero. The first term of (37.3) is therefore to be used and we can pass immediately to (26.3) with n given by (37.1) and $a_i=\tfrac{1}{2}cc'$. We obtain

$$\dfrac{d^2l}{dt^2}+\dfrac{m\kappa^2}{\kappa c}\left\{-\dfrac{1}{2}cc'\left(-\dfrac{1}{8}\kappa c\right)+\dfrac{1}{2}(n-n')c'\right\}\sin (l-l')=0.$$

The first approximation is obtained by putting $c=c_0$, $c'=c'_0$. When there is resonance we have $n_0=n'_0$. The equation therefore reduces to

(37.4) $\qquad\qquad \dfrac{d^2l}{dt^2}+\dfrac{m\kappa^2}{16}c_0 c'_0 \sin (l-l') =0.$

Similarly

$$(37.5) \qquad \frac{d^2l'}{dt^2} + \frac{m'\kappa'^2}{16} c_o c'_o \sin (l'-l) = 0.$$

From these we deduce

$$(37.6) \qquad m'\kappa'^2 \frac{d^2l}{dt^2} + m\kappa^2 \frac{d^2l'}{dt^2} = 0.$$

$$(37.7) \qquad \frac{d^2}{dt^2}(l-l') + \frac{m\kappa^2 + m'\kappa'^2}{16} c_o c'_o \sin (l-l') = 0.$$

The last equation has the standard resonance form.

38. Since c_0, c'_0 are positive by definition, equation (37.7) shows that in resonance $l-l'$ oscillates about the value zero and that $l-l' = \pi$ is the limiting case between resonance and non-resonance. Hence with the relation $n_0 = n'_0$ we must also have $\epsilon_0 = \epsilon'_0$. Thus

The stable resonance configuration of two pendulums attached to a massive block free to move horizontally is that in which the rods are always approximately parallel to one another.

If a slight disturbance be given to the system, an oscillation (libration) defined by

$$(38.1) \qquad l-l' = \lambda \sin (pt+\lambda_0), \quad p^2 = \frac{m\kappa^2 + m'\kappa'^2}{16} c_o c'_o,$$

will be present. The combination of (38.1) with (37.6) shows that the librations of the two pendulums will have opposite phases.

Suppose the pendulums had been started from opposite sides of the vertical so that $n_0 = n'_0$, but $l-l' = \pi$. Evidently M will in this case be at rest. But any small disturbance of the system will ultimately compel $l-l'$ to pass through nearly all values between π and $-\pi$, with consequent oscillation of M. Or else resonance, as defined here, will not be present, but "beats" at very long intervals will occur, according to the nature of the disturbance. Thus the motion will be highly sensitive to small disturbing forces and considerable irregularities will occur if the pendulums be used for accurate measures of time. Near the stable case, on the other hand, the irregularities will be small, but the pendulums become in effect one unit for time measurement instead of being two separate units, as in the non-resonance case, each giving its own measure.

The condition $n_0 = n'_0$, demands that

$$\kappa\left(1 - \frac{c_0^2}{16}\right) = \kappa'\left(1 - \frac{c_0'^2}{16}\right).$$

Since c_0, c'_0 were assumed to be small, this condition demands that κ, κ', and therefore by the definitions in section 36, b, b' shall be nearly the same. It is to be noticed that small differences from equality in b, b' can be compensated in resonance by the arcs through which the pendulums swing. The masses may be quite different provided they are both small compared with M.

The variation of c is given by

$$\frac{dc}{dt} = \frac{m\kappa^2}{K} \frac{\partial\phi}{\partial l} = \tfrac{1}{2}m\kappa c'\sin(l-l').$$

Substituting for $\sin(l-l')$ from (37.8) and integrating, we obtain

$$c = c_0 - \frac{8m\kappa}{c_0(m\kappa^2 + m'\kappa'^2)} \frac{d}{dt}(l-l'),$$

which, with the use of (38.1), gives

$$c = c_0 - \frac{8m\kappa}{m\kappa^2 + m'\kappa'^2} \frac{p\lambda}{c_0} \cos(pt+\lambda_0)$$

$$= c_0 - \frac{2m\kappa}{(m\kappa^2 + m'\kappa'^2)^{\frac{1}{2}}}\left(\frac{c'_0}{c_0}\right)^{\frac{1}{2}} \lambda \cos(pt+\lambda_0).$$

A similar expression gives the variation of c'. The maximum amplitudes of c and of c' we obtain from (29.1).

It is evident that neither c_0 nor c'_0 can be very small if the approximations are to be valid. If $\kappa = \kappa'$ we have $c_0 = c'_0$ and the difficulty disappears. If however the difference between κ, κ' and the starting conditions were such that c_0 or c'_0 were zero it would be necessary to reconstruct the analysis, possibly in a manner similar to that of section 33.

39. When $b = b'$, $\mu = \mu'$, it is possible to explain without transforming the equations of motion, why the unstable case of resonance is that in which the pendulums are started from rest with equal angles on opposite sides of the vertical.

Here equation (35.4) and that obtained by interchanging x, x', can be written

$$(39.1) \qquad \frac{d^2x}{dt^2} + \kappa^2 \sin x = m\kappa^2 \cos x \frac{d^2}{dt^2}(\sin x + \sin x'),$$

$$(39.2) \qquad \frac{d^2x'}{dt^2} + \kappa'^2 \sin x' = m\kappa'^2 \cos x' \frac{d^2}{dt^2}(\sin x + \sin x').$$

A particular solution of these equations when $\kappa = \kappa^1$; is

$$x = -x', \qquad \frac{d^2x}{dt^2} + \kappa^2 \sin x = 0,$$

the case in which the phases are opposite.

Suppose that a small disturbance be given to the system. The right-hand members of (39.1), (39.2) are then of the order m times disturbance, and so extremely small. The motion of one pendulum will affect the other very little, and with different amplitudes, their periods will be different and the phase difference will tend to increase until the right-hand members become large enough to affect it.

On the other hand, with the particular solution

$$x = x', \qquad \frac{d^2x}{dt^2} + \kappa^2 \sin x = 2m\kappa^2 \cos x \frac{d^2}{dt^2}(\sin x),$$

the case in which they start on the same side of the vertical, a disturbance affects both pendulums. But the difference

between (39.1), (39.2), having the factor $\cos x - \cos x'$, which
is of the order $(x-x')(x+x')$, is always very small.

The distinction between the two cases consists in the fact
that though the effect of one pendulum on the other is very
small in the unstable case, it tends to accumulate; while in
the stable case, the limit of accumulation is sooner reached
and the effect is then reversed.

40. Suppose that (35.4) and the similar equation for x'
had been reduced to the linear form at the outset by putting
$\cos x = 1$, $\cos x' = 1$, $\sin x = x$, $\sin x' = x'$. They would have
become

$$b \frac{d^2 x}{dt^2}\left(1 - \frac{\mu}{M+\mu+\mu'}\right) + gx - \frac{\mu' b'}{M+\mu+\mu'} \frac{d^2 x'}{dt^2} = 0,$$

$$b \frac{d^2 x}{dt^2}\left(1 - \left(\frac{\mu'}{M+\mu+\mu'}\right)\right) + gx' - \frac{\mu b}{M+\mu+\mu'} \frac{d^2 x}{dt^2} = 0.$$

When $b=b'$, $\mu=\mu'$, these, by subtraction and addition,
may be written

$$b \frac{d^2}{dt^2}(x-x') + g(x-x') = 0,$$

$$b\left(1 - \frac{2\mu}{M+2\mu}\right) \frac{d^2}{dt^2}(x+x') + g(x+x') = 0,$$

giving to the oscillations of $x-x'$, $x+x'$, the frequencies

$$\left(\frac{g}{b}\right)^{\frac{1}{2}}, \quad \left(\frac{g}{b} \frac{M+2\mu}{M}\right)^{\frac{1}{2}}.$$

These show the possible existence of resonance but give
no information as to the nature of the motion under such
a condition.

41. *Pendulums Mounted on a Massive Pier.* Let us now
suppose that the bar M, instead of being free, is confined
in its motion by stiff springs, so that its natural free period
of oscillation is very short compared with those of the pen-

dulums. If N be its natural frequency, the equation for the motion of M will be

$$M\left(\frac{d^2y}{dt^2} + N^2y\right) = T \sin x + T' \sin x',$$

where T, T' are the tensions of the pendulum rods.

With the earlier hypotheses concerning the magnitudes of the masses and the angles we may put $T = \mu g$, $T' = \mu' g$, $\sin x = x$, $\sin x' = x'$, in this equation so that it reduces to

$$\frac{d^2y}{dt^2} + N^2y = \frac{g}{M}(\mu x + \mu' x').$$

For oscillations of x, x' with periods very long compared with $2\pi/N$, we have N^2y large compared with d^2y/dt^2. Hence, approximately,

$$y = \frac{g}{MN^2}(\mu x + \mu' x').$$

This equation is similar to (35.1) when the latter is reduced by putting $\sin x = x$, $\sin x' = x'$, and the previous developments can be utilized by proper choices of m, m'.

The difference of chief importance is due to the fact that m, m' now have signs opposite to those of the earlier problem. Hence a reference to (37.8) shows that $l - l'$ now oscillates about the value π instead of about zero.

The present problem is substantially the same as that of two free pendulums attached to a massive pier. Such a pier will, in general, have natural periods of oscillation very short compared with those of the pendulums. Hence

The stable resonance case of two pendulums attached to a massive pier capable of oscillation is that in which the phases are opposite so that the pier does not sensibly vibrate. A small disturbance from this configuration will produce only small differences of phase.

Two Pendulums

The case separating this from the earlier problem is evidently that in which the period of oscillation of M is nearly the same as those of the pendulums. Such a case would give rise to resonances between three vibrating systems—a complicated problem which I have not attempted to attack.

V

SOLUTIONS OF $\dfrac{d^2x}{dt^2} + \kappa^2 x = m\phi'(x, t)$.

42. When $m = 0$, the solution is

$$x = c \cos(\kappa t + \epsilon),$$

so that the frequency κ is independent of the amplitude c and phase ϵ. Since the existence of the standard resonance form developed above depended on the non-vanishing of dn/dc, that theory is not applicable to the case $f'(x) = \kappa^2 x$. If ϕ' contains a periodic term with frequency κ, we have seen in section 9 that the solution contains a term whose amplitude increases continually with the time. Since infinite amplitudes do not occur in actual mechanical problems, some discussion is necessary.

In certain problems, ϕ' contains a non-linear portion $\psi'(x)$, independent of t. If this be included in $f'(x)$, the solution with the remaining portion of ϕ' neglected has no longer a constant frequency, so that dn/dc is no longer zero. It evidently contains the factor m, however, and the presence of this factor invalidates the approximate solution given in section 26, where terms factored by m^2 were neglected. No general theory for these cases appears to be at hand. Those which actually arise can usually be shown to give finite oscillations, except in quite special cases.

Another set of problems is included in the cases in which ϕ' contains terms of the form $\kappa^2 x \psi(t)$, where $\psi(t)$ is a periodic function of t. If these be included in the right-hand member, the discussion starts with the solution of

$$\frac{d^2x}{dt^2} + \kappa^2 x \{1 - m\psi(t)\} = 0,$$

a well-known type, occurring frequently in the problems of celestial mechanics. When ϕ' contains further terms which have resonance relations with those in the solution of this equation, the problems become very difficult and quite outside the limits of these lectures. They are mentioned here mainly because those which have been discussed do not give infinite amplitudes in general.

43. There are, however, certain mechanical problems which are usually classified under the equations of section 42 but to which the preceding theory does appear to be applicable. Consider, for example, the oscillations of a tightly stretched wire, under an external force which has nearly the same frequency as the principal natural frequency of the wire. Our actual experience shows that the amplitudes of the oscillations are so small that one hesitates to invoke the change of period with change of amplitude which maintains finite oscillations in the foregoing resonance theory. Further, such wires are not usually stretched in the motion sufficiently to enable us to assume a non-linear law of extensibility. Frictional damping does not appear to be a complete explanation, since it is possible to suggest experimental conditions in which this could be made very small.

The answer probably lies in a change of "natural" period due to another cause. Under the high tensions which are momentarily produced when the amplitudes are near their maxima, the framework to which the wire is attached yields to some extent. It therefore becomes part of the oscillating system and thus provides the change of period with change of amplitude which the foregoing theory shows is sufficient to prevent the development of large amplitudes. The same explanation would seem to be applicable to the oscillations under resonance conditions of any member of a frame which is under high tension when at rest.

It is known that too great rigidity of a frame is to be avoided if the frame is to be subjected to external forces which have the same periods as any combination of members of the frame. If the chief reason for this avoidance is the danger of producing additional strains due to large oscillations under resonance conditions, the foregoing theory suggests methods of preventing these additional strains other than those usually adopted. What is needed is a change in the "natural" frequency as soon as the amplitude of oscillation begins to increase. For example, if, instead of a single wire connecting two portions of a frame, we have three wires at *different* tensions and nearly in contact with one another, any increase of amplitude of any one of them will bring it into contact with one or both of the other two, causing a change of period which either destroys the resonance relation or alters the phase sufficiently rapidly so that a large amplitude cannot be built up. It is necessary that they be able to vibrate independently through very small amplitudes; if fastened together along their lengths in any way, they become equivalent to a single vibrating system with a new period of vibration which is subject to resonance in the same manner as a single wire. The same principle can evidently be applied to rods or plates. Another method is a device which shall substantially add to the mass of the vibrating body when the amplitudes exceed a certain amount, since this addition changes the natural period. These suggestions involve, not the avoidance of the resonance, but the control of the amplitude under resonance conditions.

VI

THE EFFECT OF FRICTION

44. Terrestrial mechanisms are subject to damping forces of a frictional character some of which can be represented approximately as functions of the velocity. Since such forces change sign with the velocity, it is necessary to assume that the frictional force be an odd function of the velocity when we are dealing with vibratory motion, if we are to avoid the mathematical difficulties caused by the presence of a constant which may change its sign during the motion. One example of the combined effect of friction and resonance will be given.

Consider the equation

$$(44.1) \qquad \frac{d^2y}{dt^2} = -\mu \frac{dy}{dt} - \kappa^2 \sin(y - n't - \epsilon').$$

This may be visualized as the equation of motion of a wheel subjected to a periodic couple and also to a frictional couple proportional to the angular velocity dy/dt.

Suppose that the wheel is making complete revolutions and that its angular velocity is always greater than n'. Put

$$(44.2) \qquad y = n't + \epsilon' + x,$$

so that the equation can be written

$$(44.3) \qquad \frac{d^2x}{dt^2} + \kappa^2 \sin x = -\mu \left(n' + \frac{dx}{dt} \right).$$

According to the hypothesis just made, dx/dt never vanishes and consequently (44.3) may be regarded as the equation of motion of a pendulum making complete revolutions but subjected to a disturbing force represented by the right-hand member of (44.3).

From section 18, the solution when $\mu=0$ is

$$(44.4) \qquad x = l + \frac{\kappa^2}{n^2} \sin l + \cdots, \quad l = nt + \epsilon, \quad n = p - n',$$

where p is the mean angular velocity of the wheel, that is, the mean value of dy/dt.

To obtain the solution of (44.3) when μ is not 0, we adopt the method of changing the variable, replacing x by l, n. According to the results of section 24, the equations for finding l, n are

$$(44.5) \qquad \begin{aligned} \frac{dn}{dt} &= -\frac{\mu}{K}\frac{\partial x}{\partial l}\left(n' + n\frac{\partial x}{\partial l}\right), \\ \frac{dl}{dt} &= n + \frac{\mu}{K}\frac{\partial x}{\partial n}\left(n' + n\frac{\partial x}{\partial l}\right), \end{aligned}$$

in which the right-hand member of (44.3) has been replaced by $n' + n\partial x/\partial l$, according to (24.2). The value of K is easily calculated if we follow the method used in section 31, and is found to be $K = 1 - \kappa^4/n^4 + \cdots$.

By hypothesis dx/dt, and therefore $\partial x/\partial l$, is always positive. The expression for dn/dt is therefore always negative and n is a decreasing variable. It follows that the periodic portion of x given in (44.4) has a continually increasing amplitude.

The effect of the friction is thus to decrease the mean angular velocity of the wheel but to increase the amplitude of the oscillation about this mean.

45. This motion continues until n ($=p-n'$) becomes so small that the solution (44.4) is no longer valid. In accordance with our earlier discussion on the motion of a pendulum, the discontinuous case is reached. One of two events may

happen: either x changes sign and n goes on decreasing until we reach the periodic solution

$$y = \text{periodic function of } n't + \epsilon',$$

which can satisfy (44.1). The wheel is then oscillating to and fro under the periodic force and the friction: the energy supplied by the force counterbalances that lost by friction.

Or else x begins to oscillate. In this case we have to change to the variables c, l, used for the motion of an oscillating pendulum. As in section 31 we have for the change

$$x = c \sin l + \cdots , \quad n = \kappa \left(1 - \frac{c^2}{16} + \cdots \right),$$

$$K = \kappa c \left(1 - \frac{c^2}{8} + \cdots \right).$$

The equation for c is therefore

$$\frac{dc}{dt} = -\mu \frac{(n' + cn \cos l + \cdots)(c \cos l + \cdots)}{\kappa c \left(1 - \frac{c^2}{8} + \cdots \right)}.$$

The principal part of the non-periodic term in this expression is $-\frac{1}{2}\mu c$, so that c is a variable which is oscillating but whose mean value is decreasing. It follows that x is an oscillating variable with a decreasing amplitude.

When the amplitude of x becomes small it is convenient to return to (44.3). This equation has a particular solution $x = x_0$ where

$$\sin x_0 = - \frac{\mu n'}{\kappa^2},$$

provided $\mu n' < \kappa^2$.

This solution is evidently the limiting case. The wheel is making complete revolutions with the angular velocity n' in resonance with the disturbing force, the energy supplied by the force being exactly counterbalanced by that lost in friction.

46. In order to see how this limiting case is approached, put $x = x_0 + x_1$ in (44.3), neglecting squares of x_1. We obtain

(46.1) $$\frac{d^2x_1}{dt^2} + \mu \frac{dx}{dt} + \kappa^2 \cos x_0 \cdot x_1 = 0.$$

The solution of this equation is

(46.2) $$x_1 = Ae^{\lambda_1 t} + Be^{\lambda_2 t},$$

where λ_1, λ_2 are the roots of

$$\lambda^2 + \mu\lambda + \kappa^2 \cos x_0 = 0,$$

so that

(46.3) $$\lambda_1, \quad \lambda_2 = -\tfrac{1}{2}\mu \pm (\tfrac{1}{4}\mu^2 - \kappa^2 \cos x_0)^{\frac{1}{2}}.$$

There are three cases:—

(i) $\kappa^2 \cos x_0 > \tfrac{1}{4}\mu^2$. The square root in λ_1, λ_2 is imaginary and the solution has the form

$$x_1 = Ce^{-\frac{1}{2}\mu t}\sin(qt + C_0), \quad q^2 = \kappa^2 \cos x_0 - \tfrac{1}{4}\mu^2 n'^2.$$

(ii) $\tfrac{1}{4}\mu^2 > \kappa \cos x_0 > 0$. Both roots are real and negative and the solution has the form (46.2) with $\lambda_1 < 0$, $\lambda_2 < 0$.

(iii) $\cos x_0 < 0$. Both roots are real but one is positive and the other negative so that in (46.2), $\lambda_1 > 0$, $\lambda_2 < 0$.

In case (i), x_1 approaches zero by oscillations with a decreasing amplitude. In case (ii) the approach to zero is continuous. Case (iii) appears to refer to the unstable solution, namely, to that value of x_0, which is numerically greater than $\pi/2$ in the solution of $\sin x_0 = -\mu n'/\kappa^2$.

If $\mu n' > \kappa^2$, the solution $x = x_0$, where x_0 is a constant, does not exist. There may, however, be solutions with period $2\pi/n'$ which do exist. The interest in these cases is, however, mainly mathematical and they will not be further discussed.

The principal feature of the problem is the fact that the resonance case is necessarily reached so that the oscillations have their maximum amplitude. After this stage is passed the amplitude diminishes and, dependent on the initial conditions, one of the types of steady motion is the final outcome. Theoretically these types only exist as t approaches infinity. It should be pointed out, however, that with oscillations of very small amplitude the so-called "statical" friction, which has been neglected, becomes important and actually brings the system to rest or relative rest in a finite time.

For EU product safety concerns, contact us at Calle de José Abascal, 56–1°,
28003 Madrid, Spain or eugpsr@cambridge.org.